LABORATORY MANUAL
NATURAL DISASTERS
REVISED PRINTING

Kenneth Weide

Edited by
John Oughton

Kendall Hunt
publishing company

Cover image © Shutterstock, Inc.

Kendall Hunt
publishing company

www.kendallhunt.com
Send all inquiries to:
4050 Westmark Drive
Dubuque, IA 52004-1840

Copyright © 2015 by Kendall Hunt Publishing Company
Revised Printing 2017

ISBN 978-1-5249-1420-2

All rights reserved. No part of this publication may be reproduced,
stored in a retrieval system, or transmitted, in any form or by any means,
electronic, mechanical, photocopying, recording, or otherwise,
without the prior written permission of the copyright owner.

Published in the United States of America

I would like to thank the administration of Century Community College, White Bear Lake, Minnesota for the opportunity and support to create this laboratory manual. Also, Roger Carlson from Hamline University, St. Paul, Minnesota for his ideas for Exercise 2, Part 2.

Contents

Exercise 1 **Geographic Location and Natural Disasters** 1
What is Latitude? 1
What is Longitude? 2

Exercise 2 **Plate Tectonics and Volcanic Hazards** 7
Plate Tectonics 7
Volcanic Hazards 9

Exercise 3 **Worldwide Earthquake Activity and Distribution** 11

Exercise 4 **Determining Earthquake Location and Magnitude** 15

Exercise 5 **Sun Angles and Solar Radiation** 23
The Analemma 23
Equinox (Equal Night) 25
Solstice (Sun Stops) 25
Arctic and Antarctic Circles (66 1/2 degrees N and S) 26
Sun Angles 26

Exercise 6 **Severe Weather and Weather Mapping** 29
Air Pressure 31
Wind Speed and Direction 31
Temperature and Dew Point 31
Front Locations 32
Warm Front 34
Cold Front 34
Dry Line 34

Exercise 7	**Hurricane Tracking** 35	
	Hurricane Andrew, Florida and the Gulf of Mexico, 1993 35	
Exercise 8	**Climate Change** 39	
	Minneapolis/St. Paul Climate 39	
Exercise 9	**Acid Precipitation Across the Nation** 41	
Exercise 10	**Topographic Map Interpretation** 45	
	Direction 45	
	Scale and Distance 45	
	Elevation 46	
	Contour Line Patterns 46	
Exercise 11	**Regional Floods** 49	
	St. Croix River, Minnesota/Wisconsin Border 49	
Exercise 12	**Mega Disasters** 51	
	Tsunami 51	
	Impact 53	

APPENDIX A 55

APPENDIX B 67

APPENDIX C 75

CREDITS 85

EXERCISE 1

Geographic Location and Natural Disasters

It is vital to know the exact location of any particular natural disaster, be it the path of a hurricane or the epicenter of an earthquake. The most common method for earth location is latitude and longitude as seen in Figure 1.1. Latitude and longitude degrees can be further broken into minutes and seconds, 60 minutes in a degree and 60 seconds in a minute. In reality this would pinpoint a location to within approximately 100 feet.

1. What is Latitude?

(a)

Figure 1.1a Equator We can imagine the Earth as a sphere, with an axis around which it spins. The ends of the axis are the North and South Poles. The Equator is a line around the Earth, an equal distance from both poles. The Equator is also the latitude line given the value of 0 degree. This means it is the starting point for measuring latitude. Latitude values indicate the angular distance between the Equator and points north or south of it on the surface of the Earth.

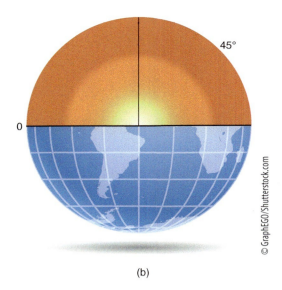

(b)

Figure 1.1b A line connecting all the points with the same latitude value is called a line of latitude. This term is usually used to refer to the lines that represent values in whole degrees. All lines of latitude are parallel to the Equator, and they are sometimes also referred to as parallels. Parallels are equally spaced. There are 90 degrees of latitude going north from the Equator, and the North Pole is at 90 degrees N. There are 90 degrees to the south of the Equator, and the South Pole is at 90 degrees S. When the directional designators are omitted, northern latitudes are given positive values and southern latitudes are given negative values.

2. What is Longitude?

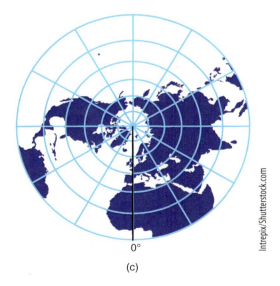

(c)

Figure 1.1c **Prime Meridian** Lines of longitude, called meridians, run perpendicular to lines of latitude, and all pass through both poles. Each longitude line is part of a great circle. There is no obvious 0-degree point for longitude, as there is for latitude. Throughout history many different starting points have been used to measure longitude. By international agreement, the meridian line through Greenwich, England, is currently given the value of 0 degree of longitude; this meridian is referred to as the Prime Meridian. Longitude values indicate the angular distance between the Prime Meridian and points east or west of it on the surface of the Earth.

In this exercise you will locate regions on the Earth that are prone to natural disasters and sites of major disasters throughout history. Using an atlas and the map provided in Appendix A, *number and label* the following locations, including the latitude and longitude.

1. Bangladesh 1970. Cyclone ravaged then East Pakistan on the coast of the Bay of Bengal. Fatalities 400,000 est.

    Lat. _____   Long. _____

2. Valdivia, Chile 1960. Greatest earthquake recorded, 1960, mag. 9.5.

    Lat. _____   Long. _____

3. Sumatra 2004. Earthquake 9.3, with accompanying tsunami. Fatalities 245,000 est.

    Lat. _____   Long. _____

4. Prince William Sound, 1964. Mag. 9.2, largest earthquake recorded in North America.

    Lat. _____   Long. _____

5. Cape Verde Islands. Tropical storm formation area.

    Lat. _____   Long. _____

6. Sicily. Mt. Etna, Europe's most active volcano.

    Lat. _____   Long. _____

7. Shaanxi, China, 1556. Deadliest known earthquake. Fatalities 830,000 est.

    Lat. _____   Long. _____

8. Yangtze River, 1937. Deadliest flood on record. Fatalities unknown, 1–3,700,000 est.

    Lat. _____   Long. _____

9. Philippine Islands. Mt. Pinatubo eruption, 1991, largest in last 100 years.

    Lat. _____   Long. _____

10. Iceland. Tectonic spreading center volcanism.

    Lat. _____   Long. _____

EXERCISE 1: GEOGRAPHIC LOCATION AND NATURAL DISASTERS

11. Cascade Range, Mt. St. Helens. Active North American volcanism.

    Lat. _____ Long. _____

12. Martinique, Mt. Pelee, 1902. Fatalities 30,000 est.

    Lat. _____ Long. _____

13. Armero, Colombia, 1985. Volcanic lahar, 22,000 buried.

    Lat. _____ Long. _____

14. Cameroon, 1986, Lake gas, CO_2 eruption. Fatalities 1,700 est.

    Lat. _____ Long. _____

15. Sri Lanka, 2004, tsunami. Fatalities 30,000 est.

    Lat. _____ Long. _____

16. Yucatan Peninsula. Asteroid impact, 64 million BC. Suspected dinosaur extinction.

    Lat. _____ Long. _____

17. Andes Mountains. Active South American volcanoes.

    Lat. _____ Long. _____

18. Haiti 2010 earthquake, mag. 7.0. Fatalities 230,000 est.

    Lat. _____ Long. _____

19. New Madrid, Missouri. Mid-continent rift. Series of earthquakes, 1811–1812. Changed the course of the Mississippi River.

    Lat. _____ Long. _____

20. Tonga Trench. Tectonic subduction zone, volcanic islands, and earthquakes.

    Lat. _____ Long. _____

21. Aleutian Trench. Tectonic subduction zone volcanism and earthquakes.

    Lat. _____ Long. _____

22. Hawaii. Active volcanism.

   Lat. _____ Long. _____

23. Krakatoa, 1883. Volcanic island explosion and tsunami. Fatalities 36,000 est.

   Lat. _____ Long. _____

24. Mexico City 1985. Earthquake mag. 8.1. Fatalities 10,000 est.

   Lat. _____ Long. _____

25. Kobe, Japan, 1995. Earthquake mag. 6.9. Fatalities 5,500 est.

   Lat. _____ Long. _____

26. Yellowstone. Super volcano. Three major eruptions over last two million years.

   Lat. _____ Long. _____

27. Italy, Mt. Vesuvius, AD 79 volcanic eruption. Fatalities unknown, est. in the 1,000s.

   Lat. _____ Long. _____

28. Himalaya Mts., tectonic convergence zone, earthquakes.

   Lat. _____ Long. _____

29. Ethiopia, 1984–1985. Drought and famine. Fatalities unknown, 100,000s est.

   Lat. _____ Long. _____

30. Marianas Trench. World's deepest trench. Tectonic subduction zone volcanism and earthquakes.

   Lat. _____ Long. _____

EXERCISE 2

Plate Tectonics and Volcanic Hazards

1. Plate Tectonics

Plate tectonic theory is a relatively new geologic paradigm developed during the 1960s. According to this theory, the Earth's crust is fractured into huge slabs, or plates, that are made of oceanic and continental rock. Figure 2.1 shows the different processes that take place. At divergent plate boundaries molten magma wells up to create new crust. Because of density differences between plates, where converging plates collide, one dives below the other, back into the mantle, to be recycled in a process called subduction. At these margins, active volcanism occurs, accompanied by numerous earthquakes. Such is the case along the coast of the Pacific Northwest of the United States (Figure 2.2).

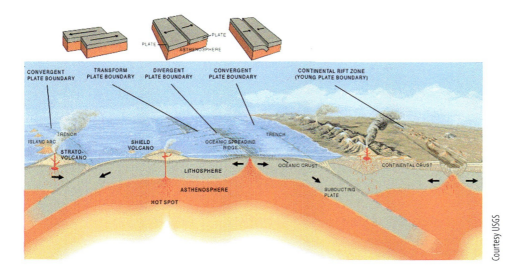

Figure 2.1 Plate tectonic processes.

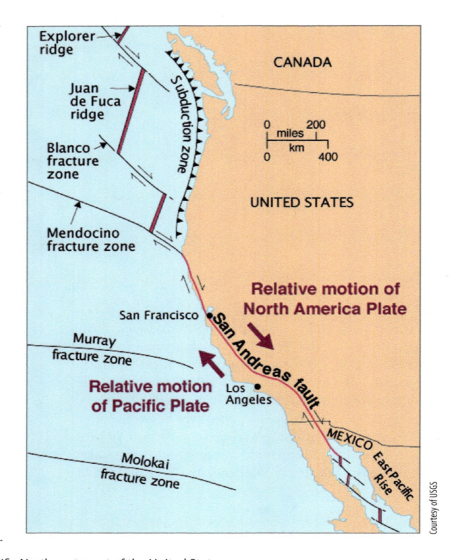

Figure 2.2 Pacific Northwest coast of the United States.

A remnant of a once large oceanic plate is diving below the North American plate creating the volcanic Cascade mountain range. The Juan de Fuca plate is moving eastward at an average rate of 3–4 cm/y United States Geological Survey (USGS).

1. At the rate of 3.5 cm/y, how long will it take for the entire plate to subduct under the North American plate? A conversion table can be found in Appendix B.

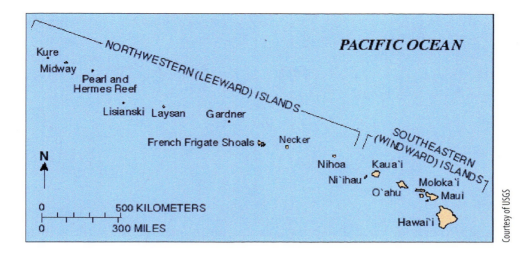

Figure 2.3 Map of the Hawaiian Islands, a chain of volcanoes that stretches about 2,700 km in a northwesterly direction from the Island of Hawaii. The age of the volcanoes that form the islands increases progressively from Hawaii, where the extinct volcanoes are about 30 million years old at the northwest end. The chain of volcanoes continues as seamounts for another 3,000 km; the chain bends sharply to the northward about 700 km beyond this map and becomes the Emperor Seamounts.

The Hawaiian Islands formed from a different process than subduction volcanism. They formed as the Pacific plate has moved over a tectonic hot spot, or stationary mantle plume, that has melted through the crust. As the plate continues to move, it eventually disconnects from an island to start a new island from the sea floor upward. This process has formed islands that extend thousands of kilometers to the northwest (Figure 2.3).

2. What has been the average rate of plate movement in cm/y from Kure Island at the northwest end of the chain, to Hawaii at the southeast? Measure from the center of Hawaii.

2. Volcanic Hazards

Volcanism at spreading centers and hot spots produces basaltic lavas of low viscosity compared with lavas at subduction zones. At subduction zones, lavas are andesitic that are silica rich, increasing their viscosity. Highly viscous lavas tend to be explosive, releasing pent up gases near the surface once pressure has been reduced.

Most fatalities from volcanic eruptions occur near subduction zone volcanoes, either from the hot ash clouds called pyroclastic flows that roar down the mountain side at hundreds of miles per hour, or glacial melt water mixing with the ash, forming mud flows that follow river channels away from mountains. These mud flows, called lahars, can travel at speeds of 60 kph, and have buried river towns tens of kilometers away from a volcano.

You will use Mt. Rainier in Washington State to determine the amount of warning nearby towns might have if a lahar threatens them. Mt. Rainier is currently dormant but has erupted historically and undoubtedly will again (USGS). A lahar and pyroclastic flow hazard map developed by the USGS, found in Appendix A, will be used for this exercise.

1. Many towns lie in the river valleys leading away from Mt. Rainier. Two rivers, the Carbon and the Puyallup, flow to the northwest from Mt. Rainier toward Tacoma and Seattle. The first town to be affected by lahar would be the small town of Carbonado, population 600. What is the straight line distance in kilometers to Carbonado from where the Carbon River first intersects the park boundary?

2. Rivers do not flow in a straight line very long and will change course and meander. If we use the straight line distance calculated in #1, as the minimum amount of travel distance for a lahar, how much time will elapse before it reaches Carbonado if traveling at 60 kph?

3. The Carbon River flows into the Puyallup River just southeast of Orting, a town with a population of 3,800. This could produce an even bigger lahar. How long would it take a lahar to reach Orting flowing down the Puyallup, measured from where the north branch of the Puyallup leaves the park? Again, just measure the straight line distance.

4. The even larger town of Sumner, pop. 8,500, lies a short distance to the north. How long will the lahar take to reach it from Orting?

5. The travel times calculated would be an underestimate for a flow rate of 60 kph. Even at half again as long, accounting for a nonlinear flow of the lahar, the travel times are alarmingly short. What complicating factors might produce even greater casualties other than the short travel time of the lahars?

EXERCISE 3

Worldwide Earthquake Activity and Distribution

This exercise will utilize the internet to gather data on global earthquake activity over a two-week period. You will be using a US government website at http:earthquake.usgs.gov. Scroll down to the bottom left and click Significant Earthquake Archive. Use the previous years data. Print this out for future use.

The data will be organized by region, number, and magnitude of earthquakes. Magnitude refers to the relative size of an earthquake. The first, and still popular earthquake measurement technique is the Richter scale, developed in the 1930s in California. It is called a local scale because its accuracy decreases over distance (other scales have since been constructed that alleviate this problem). For each increase in magnitude, the size of the earthquake increases tenfold, measured by the seismic wave amplitude trace. However, the energy release increases thirty times. For example, a magnitude 6.0 quake releases 30×30 the energy than a magnitude 4.0. Your data will be organized on Table 3.2, similar to that in Table 3.1.

One idea employed to try to predict the locations of future earthquakes is the seismic gap method. Simply, if an area along a major fault, such as the San Andreas in California has not had a significant quake recently, a quake may be more likely to occur than at other areas that have experienced quakes.

EXERCISE 3: WORLDWIDE EARTHQUAKE ACTIVITY AND DISTRIBUTION

Table 3.1 Significant Earthquakes Worldwide in 2013

Region or Country	Number of Quakes	Average Magnitude
Aleutian Islands/Alaska	3	7.0
Caribbean Region	0	
Central America	3	6.1
China	2	6.3
Europe/Mediterranean	3	6.1
India/Mid-east	5	6.9
Indonesia/Philippines	20	9
Japan	11	6.5
Mexico	4	5.9
South Atlantic	5	7.1
South Pacific	5	6.7
Southern United States	3	4.4
Western South America	5	6.8
Western United States	9	4.6
Other	6	4.1

Source: United States Geological Survey

Table 3.2 Earthquakes Worldwide Year _____

Region or Country	Number of Quakes	Average Magnitude
Aleutian Islands/Alaska		
Caribbean Region		
Central America		
China		
Europe/Mediterranean		
India/Mid-east		
Indonesia/Philippines		
Japan		
Mexico		
South Atlantic		
South Pacific		
Southern United States		
Western South America		
Western United States		
Other		

1. The data in Table 3.1 shows three areas with more significant earthquakes than the others; Indonesia, the Philippines, and the Western United States. Does your data show a similar pattern of earthquake frequency for these regions? If not, which regions differ?

2. Why do you think the regions in Table 3.1 have the highest frequency of significant earthquakes? Do they have something in common tectonically?

3. The Caribbean region in 2013 had no significant quakes even though it has experienced devastating quakes in the past evidenced by the 2010 Haiti event. Using the seismic gap method on a global scale, the Caribbean may be due for a significant quake in the near future. What regions in your data may be candidates for a significant earthquake?

Earthquakes are caused by crustal movement along faults, magma movement in volcanic regions, and some think by new drilling techniques.

4. The *Other* category in the 2013 data contains six quakes, two in central Russia and one in Canada. These continental interior quakes are probably along old fault lines. Two occurred on Hawaii. What may have caused these Hawaiian quakes? Another quake, the strongest of the six, at magnitude 6.6, occurred in the North Atlantic. What may have caused this quake? The Southern United States quakes?

Determining Earthquake Location and Magnitude

EXERCISE 4

California is one of the most active earthquake regions in the world, with 20–30 occurring on a daily basis. Of course most of these are relatively small, causing little or no damage, and often are not even felt. Earthquakes are caused by magma movement, common in volcanic regions, or movement along weak areas of the bedrock called faults. Your task will be to determine the magnitude and location, or epicenter, of a sizeable earthquake that occurred in California in 2004.

The nomogram in Figure 4.1 is used to calculate Richter scale magnitude using the difference in P and S wave arrival times at a seismic station, and the amplitude of the seismic wave trace. Primary, or P waves, are the fastest and therefore the first to arrive at a seismic station. Secondary, or S waves, follow behind, and leave a larger seismic wave trace. The difference in arrival time of the two waves is a function of distance. The size, or amplitude of the wave trace, is measured in millimeters. Use the seismograms in Figures 4.2, 4.3, 4.4 to calculate the distance from the epicenter and the magnitude of the earthquake using the nomogram in Figure 4.1. Use an average from the three locations for your final magnitude.

EXERCISE 4: DETERMINING EARTHQUAKE LOCATION AND MAGNITUDE

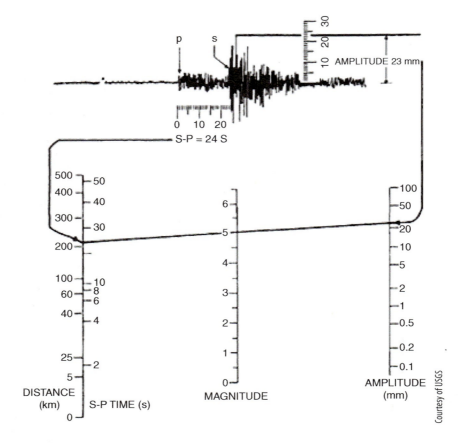

Figure 4.1 Example calculation for a magnitude 5 earthquake.

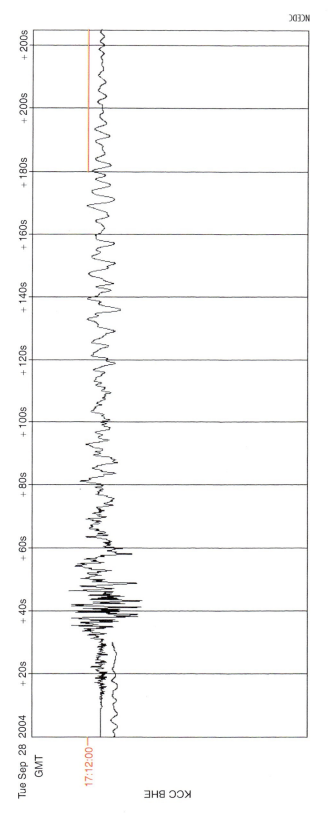

Figure 4.2 UC Berkeley Seismogram for Station KCC.

1. S–P wave _____ Wave Amplitude _____ Magnitude _____

EXERCISE 4: DETERMINING EARTHQUAKE LOCATION AND MAGNITUDE

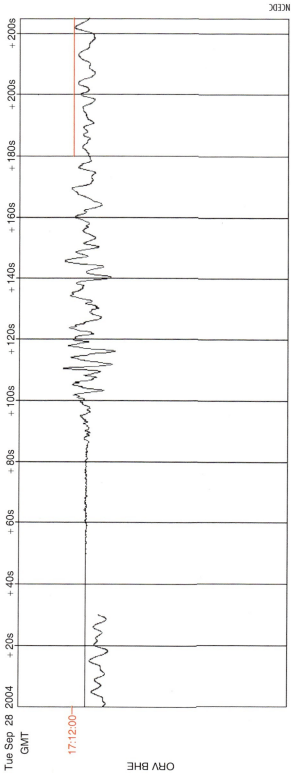

Figure 4.3 UC Berkeley Seismogram for Station ORV.

2. S–P wave _____ Wave Amplitude _____ Magnitude _____

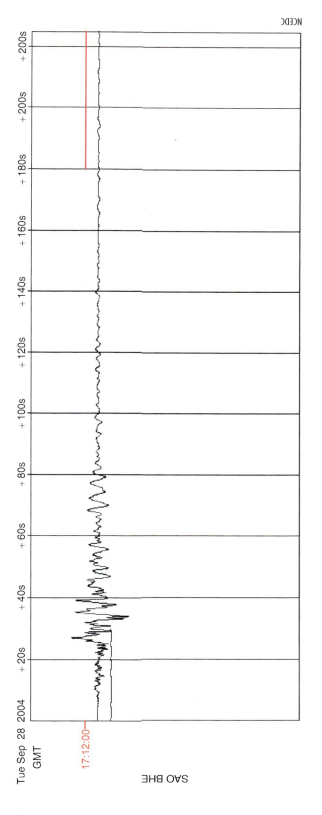

Figure 4.4 UC Berkeley Seismogram Station SAO.

3. S–P wave _____ Wave Amplitude _____ Magnitude _____

EXERCISE 4: DETERMINING EARTHQUAKE LOCATION AND MAGNITUDE

4. Average Magnitude for the three seismograms _____

5. Using the Mercalli scale of earthquake intensity in your textbook, would you classify this quake as minor, moderate, or major?

Rapid determination of the epicenter of an earthquake enables rescue personnel to respond as fast as possible to areas most damaged. In an earthquake, an hour lost in the initial response may very well make the situation worse, and most likely will.

Three seismograms are needed to calculate an earthquakes epicenter. The epicenter marks the point at the Earth's surface directly above where the energy was released. Drawing radial distance circles already calculated for the three stations, will produce an intersection point corresponding to the approximate epicenter like in Figure 4.5.

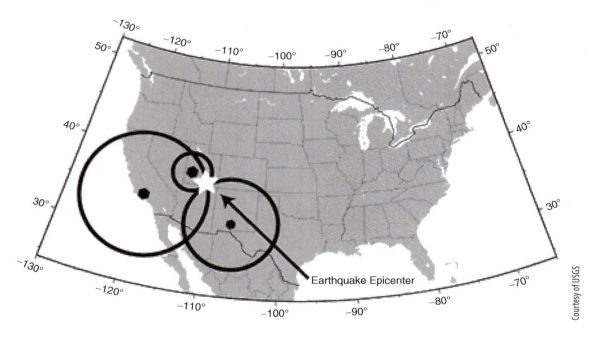

Figure 4.5 Radial triangulation.

The map of California in Figure 4.6 will be used for your calculation. Locate the three stations, and using the map scale provided, draw a radial circle for each station with a pencil compass.

Epicenter Lat. _____ Long. _____ Nearest Station _____

6. Is this location a rural or suburban area?

Figure 4.6 Map of California seismic stations. This map can be found at: http://seismo.berkeley.edu/annual_report/ar01_02/node7.html

7. Depending on the location and magnitude, in your opinion, did this quake cause extensive damage and loss of life?

EXERCISE 5

Sun Angles and Solar Radiation

External energy receipt from solar radiation is thousands of times more than the internal energy of the Earth. This energy drives the weather and the associated phenomena that develop. The more vertical the sun's rays are reaching the Earth, the more intense the radiation is per unit area. Where you live on the Earth will determine how direct the solar radiation will be. For any given day of the year, there is only one latitude where the noon sun will be directly overhead.

1. The Analemma

The angle of the noon sun off of the horizon is called the elevation angle of the sun. This angle varies throughout the year for any given latitude due to the tilt of the Earth's axis. As the Earth is tilted 23 1/2 degrees from the orbital plane around the sun, the sun will be directly overhead, at noon, between 23 1/2 degrees N latitude and 23 1/2 degrees S latitude on any day of the year.

The latitude of the overhead noon sun is termed declination. It can be determined by using the analemma (Figure 5.1). Find the date in question and read off to the left for the latitude (declination) of the overhead sun.

Where is the noon sun overhead for the following dates:

1. April 15 _____
2. March 22 _____
3. June 22 _____
4. October 10 _____
5. August 9 _____
6. September 22 _____
7. December 22 _____
8. November 25 _____

24 EXERCISE 5: SUN ANGLES AND SOLAR RADIATION

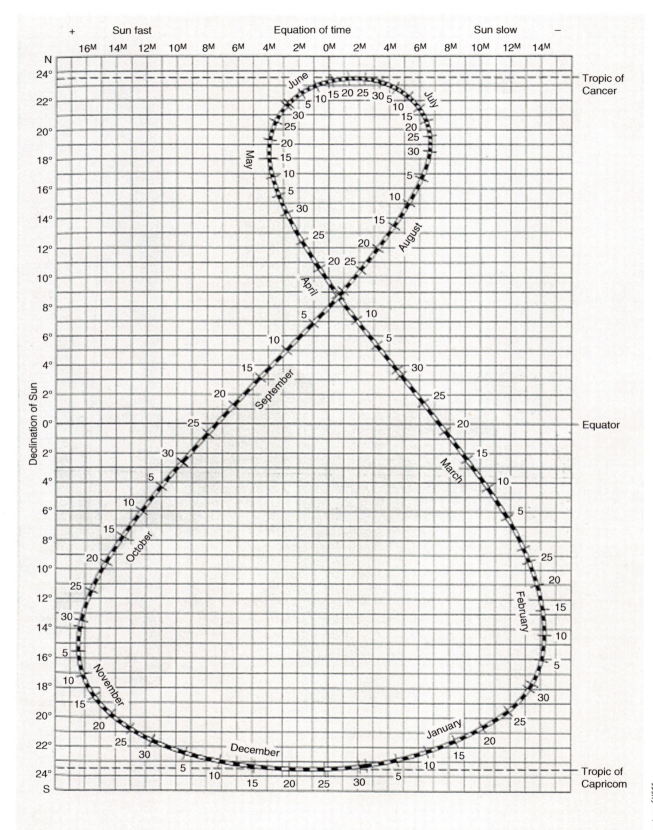

Figure 5.1 The analemma.

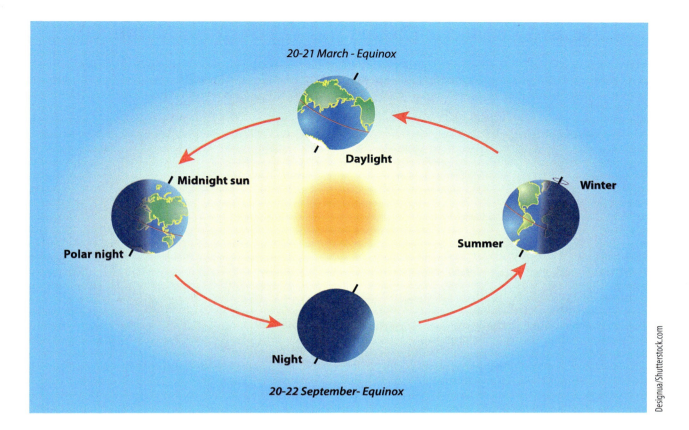

Figure 5.2 Equinox and solstice.

2. Equinox (Equal Night)

The solar equinox occurs twice a year on March 22 and September 22 (Figure 5.2). This is when the noon sun is directly overhead at the equator (0 degree latitude). On these dates, day-time and night-time hours are the same everywhere on the Earth, twelve hours of each. (Note: the equator is the pivotal point throughout the year, and is the only latitude that always has equal day and night).

March 22 marks the first day of spring for the northern hemisphere and is called the Vernal equinox, and is the first day of fall in the southern hemisphere called the Autumnal equinox. September 22 marks the first day of fall for the northern hemisphere and the first day of spring for the southern hemisphere.

3. Solstice (Sun Stops)

The solstice refers to the farthest north and south march of the overhead noon sun, corresponding to 23 1/2 degrees N (Tropic of Cancer), and 23 1/2 degrees S (Tropic of Capricorn). This occurs on June 22 and December 22. The difference between day-light and night-time hours is at their greatest (except for the equator). These dates are often called the longest or shortest days of the year.

June 22 marks the first day of summer for the northern hemisphere (summer solstice) and the first day of winter for the southern hemisphere (winter solstice). December 22 marks the first day of winter for the northern hemisphere and the first day of summer for the southern hemisphere.

4. Arctic and Antarctic Circles (66 1/2 degrees N and S)

Ninety degrees of arc on a circle is analogous to a 90-degree right angle. More than 90 degrees of arc is like going around a corner. You can't see around it. On the Earth, if you live more than 90 degrees away from where the sun is overhead at noon, the sun will not rise. You may experience a dawn but no sunrise. For example, if it is December 22 the sun is directly overhead at noon at 23 1/2 degrees S. If you live beyond 66 1/2 degrees N (90 degrees of latitude away) the sun will not rise. The opposite is true beyond 66 1/2 degrees S, the sun will not set.

5. Sun Angles

Here is how to do it. When the sun is directly overhead, the angle of the sun off of the horizon is considered to be 90 degrees. For every degree of latitude you are away from where the noon sun is directly overhead, you must *subtract* that number of degrees from 90 to determine the elevation angle. For example, if the noon sun is directly overhead at 0 degree (the equator), the noon sun angle at 40 degrees N latitude is 50 degrees off of the southern horizon. At 40 degrees S latitude, the sun angle is also 50 degrees but off the northern horizon.

9. If you live at 20 degrees S latitude, off of which horizon is the sun on June 22?

   _____

10. If you live at 5 degrees S latitude, off of which horizon is the sun on December 22?

    _____

11. What is your latitude if the noon sun is 25 degrees above your southern horizon on September 22?

    _____

12. What is your latitude if the noon sun is 44 degrees above your southern horizon on November 17?

    _____

13. What is your latitude if the noon sun is 25 degrees above your northern horizon on December 1?

    _____

14. What is your latitude if the noon sun is 14 degrees above your southern horizon on January 21?

    _____

15. If the noon sun is 37 degrees off the northern horizon at 58 degrees S, what date is it?

    _____

16. How far above your horizon would the noon sun be on the 22nd of August in Minneapolis (45 degrees N)?

    _____

17. If the noon sun is 30 degrees above your southern horizon at 50 degrees N, at What latitude is the sun overhead?

    _____

18. What is the angle of the sun above the horizon at 78 degrees N latitude on December 30?

    _____

EXERCISE 6

Severe Weather and Weather Mapping

Perhaps the most feared storm in nature is the tornado, for its sudden onset, severe winds, and unpredictability. On March 29, 1998 a family of tornadoes struck southwestern Minnesota at about 4:30 PM, causing widespread destruction and fatalities. It was unique for the time of year, and for its rapid development. You will analyze this event in this exercise.

Tornadoes were called cyclones by Midwesterners in the early twentieth century. A cyclone is actually a region of low pressure that has a mappable center, such as in Figure 6.1. This map shows pressure lines called isobars at 4-millibar intervals. The pressure is constant all along the isobar. Average sea level pressure is 1,013 millibars. Less than 1,013 is considered low pressure, greater than 1,013 is high pressure. You will construct such a map in this exercise. An isobar mapping tutorial can be found in Appendix C. Note the weather data on the map for each city. This is called a station model and will be discussed later.

Severe weather outbreaks occur when there is a large contrast in temperature and moisture over a horizontal distance. The boundary between contrasting air masses is the weather front (Figure 6.2). Weather fronts are only connected to low pressure centers. They are never connected to high pressure centers. Warm fronts are typically found north and east of a low, while cold fronts are found trailing to the south and west of a low. A special type of front is called a dry line front that separates areas of moisture contrast and is usually found south of the low pressure center. Go to Appendix C to see the different types of fronts.

Weather is usually the most severe along cold fronts and dry lines. This is because that is where the greatest contrast in temperature or moisture is usually found. However, warm fronts can also produce severe weather under the right conditions. Your task will be to determine which type of front produced the March 29th tornado outbreak.

Vertical lift of air is the result of clashing air masses. Warm air, being less dense than cold air, will easily rise over cooler air at both warm and cold fronts. Rising air cools to dew point and condenses, producing precipitation and possibly severe weather.

EXERCISE 6: SEVERE WEATHER AND WEATHER MAPPING

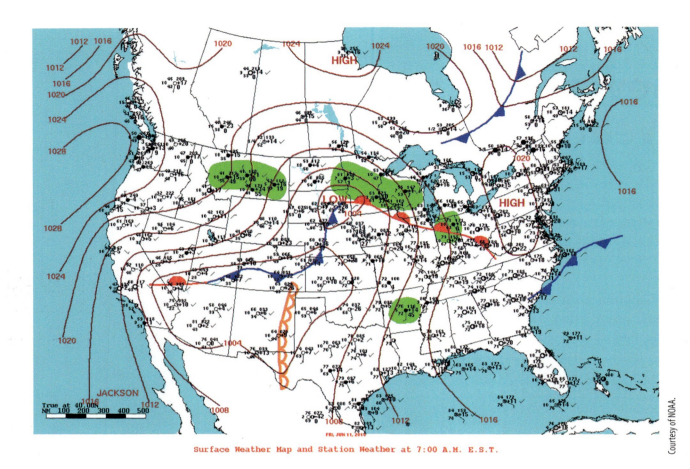

Figure 6.1 Typical weather map, June 11, 2010.

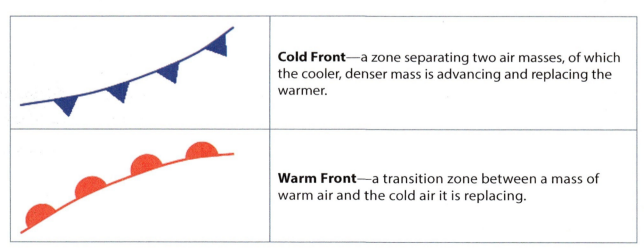

Figure 6.2 Weather front symbols.

1. You will create a weather map using the map provided of the upper Midwest in Appendix A, and the data in Table 6.1. Obtain a US road atlas to accurately locate each city. Place a dot at city locations on the Midwest map. Using the data in Table 6.1, create a station model for each city like the one in Figure 6.3 below. You will be plotting pressure, wind speed and direction, temperature, and dew point only.

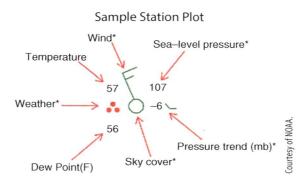

Figure 6.3 Weather station model.

1. Air Pressure

On a true station model, air pressure is abbreviated, where the 9 or10 has been omitted from the actual value to decrease clutter, with the last digit in tenths of millibars. The air pressure in Figure 6.3 would read 1,010.7 millibars. You may use the whole value for the stations given in Table 6.1, placing it to the upper right of the city dot as in Figure 6.3.

2. Wind Speed and Direction

Wind speed is recorded in knots. One knot is equal to slightly more than one mile per hour at 1.15 mph. A wind flag is used to depict both speed and direction, explained in Figure 6.4.

3. Temperature and Dew Point

Temperature and dew point are both measured in degrees Fahrenheit in this exercise. Dew point temperature is the temperature at which condensation will occur. Air that is cooled to dew point is said to have reached saturation, and condensation must take place. Dew point can also be thought of as a measure of the amount of moisture in the air, and can be quantified as so many grams of water vapor per kilogram of air. Warm air has a higher moisture capacity than cooler air. As the air temperature increases so does the moisture capacity. Dew point temperatures in the 60s and 70s are considered tropical. Temperature is plotted above dew point on the left side of the station model.

Once the station models have been placed on the map, your next task is to draw in isobars at 2-milibar intervals starting with 992 millibars, and working away from the center of the low. It is strongly suggested that you work the tutorial on pressure mapping in Appendix C first, to understand the mapping process. Next, locate the fronts of the cyclone.

4. Front Locations

Air spirals into a cyclone in a counter clockwise fashion. Because of this, air to the east of a cyclone will generally be from a southerly direction, and to the west, a northerly direction. South winds bring warm are to the north, and north winds cool air to the south. Therefore, temperature changes across a front should be noticeable. Figure 6.1 should provide a good guide.

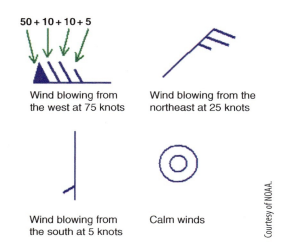

Figure 6.4 Wind speed and direction symbol. Wind is plotted in increments of 5 knots (kts), with the outer end of the symbol pointing toward the direction from which the wind is blowing. The wind speed is determined by adding up the total of flags, lines, and half-lines, each of which has the following individual values:

Flag: 50 kts
Line: 10 kts
Half-Line: 5 kts

Table 6.1   Upper Midwest 1500 h. Weather Station Data March 29, 1998

	Pressure (mb)	Wind Speed (kts)/ Direction	Temperature (F)	Dew Point (F)
Nebraska				
Alliance	1,002	NW 15	34	32
North Platte	995	NE 10	62	46
McCook	993	SW 20	70	26
O'Neill	993	NW 10	49	39
Grand Island	994	SW 20	78	38

	Pressure (mb)	Wind Speed (kts)/ Direction	Temperature (F)	Dew Point (F)
Norfolk	992	SW 25	79	39
Omaha	995	SW 20	76	54
Kansas				
Manhattan	996	S 20	77	57
Missouri				
Kansas City	1,000	S 20	76	58
Iowa				
Sioux city	994	S 25	75	63
Des Moines	998	SW 25	76	58
Algona	999	SW 15	70	64
Waterloo	997	SE 20	80	56
Dubuque	1,000	S 20	79	54
South Dakota				
Pierre	1,002	N 15	39	37
Chamberlain	1,000	NE 15	39	37
Aberdeen	1,002	NE 20	34	28
Watertown	997	NE 15	39	38
Sioux Falls	993	N 20	57	54
Minnesota				
Redwood Falls	996	NE 5	54	54
Worthington	996	SE 10	66	64
Rochester	997	SE 10	62	58
Mpls/StP	998	SE 5	53	50
St. Cloud	999	N 10	50	50
Brainerd	1,001	NE 15	41	39
Bemidji	1,002	NE 10	37	35
Fergus Falls	1,002	NE 15	39	39
Wisconsin				
Hayward	1,002	SE 10	54	47
Wausau	1,003	SE 15	52	49

NOAA.

5. Warm Front

Notice in Figure 6.1 that the warm front is located to the east of the center of the cyclone. This is typical. Wind direction north of the front is usually out of the southeast, and southwest south of the front. Also, there should be an increase in temperature and dew point to the south of the front. Remember, the greater the contrast in temperature and moisture, the greater potential for severe weather.

6. Cold Front

The cold front in Figure 6.1 is to the west of the center of the cyclone. Wind direction south of the front is typically out of the southwest, and north of the front from the northwest. Temperature and dew points should be lower behind the cold front. The cold front is often the zone of greatest contrast.

2. Using this information locate the two fronts on your map as accurately as possible.

7. Dry Line

Not every cyclone will exhibit what is called a dry line. This is a rapid change in dew point, typically increasing from west to east, south of the center of the low. This area south of the center, between the warm and cold front, is called the warm sector of the cyclone. If the warm sector contains a dry air mass originating from the desert southwest, and a moist air mass from the Gulf of Mexico is present, large contrasts in moisture content might be expected across the warm sector. Interestingly, dry air is denser than moist air, and acts as a lifting mechanism when they come into contact. Hence a dry line front is a change in moisture content horizontally, even though there is little change in air temperature.

3. Inspect the map and see if a dry line might be present. If so, locate the front on the map with a dashed line.

4. Which type of front do you think was the cause of the tornado outbreak in southwestern Minnesota?

EXERCISE 7

Hurricane Tracking

1. Hurricane Andrew, Florida and the Gulf of Mexico, 1993

Hurricane tracking and predicting landfall is not a precise science, and can only be broadcast as probability of where a hurricanes eye will come ashore. In the summer of 1993, Hurricane Andrew ravaged parts of the Bahamas, Florida, and Louisiana. It was the costliest natural disaster in US history up to that time. Your task will be to plot the course of the storm and analyze the probability data for landfall at Miami, FL and New Orleans, LA.

Atlantic Hurricane Tracking Chart

1. Table 7.1 contains the 0000 hours (midnight) and the 1200 hours (noon) position of the center of the storm in degrees latitude and longitude. Using this data, plot the position of the storm as accurately as you can every twelve hours, starting with 0000 hours on the 17th on the Atlantic Basin chart provided in Appendix A. Use a small blank circle for 0000 hours position, and a filled in circle for the 1200 hours position. Include the date for each point.

2. The scale of the Atlantic tracking chart is 1:45,000,000. Therefore, one millimeter on the chart is equal to 45 kilometers in reality. What date and time did the hurricane Andrew move the fastest?

3. Andrews speed appears constant from 23/0000 to 25/0000. Speed is calculated as distance divided by time. What was its speed in kph? Convert kilometers to miles. What is its speed in mph?

EXERCISE 7: HURRICANE TRACKING

Table 7.1 Track of Hurricane Andrew 1992

Date/Time	Position		Date/Time	Position	
	Lat.(N)	Long.(W)		Lat.(N)	Long.(W)
17/0000	11.2	35.5	22/1200	25.8	68.3
17/1200	12.3	42	23/0000	25.6	71.1
18/0000	13.6	46.2	23/1200	25.4	74.2
18/1200	14.6	49.9	24/0000	25.4	77.5
19/0000	16.3	53.5	24/1200	25.6	81.2
19/1200	18	56.9	25/0000	26.2	85
20/0000	19.8	59.3	25/1200	27.2	88.2
20/1200	21.7	60.7	26/0000	28.5	90.5
21/0000	23.2	62.4	26/1200	30.1	91.7
21/1200	24.4	64.2	27/0000	31.5	91.1
22/0000	25.3	65.9	27/1200	32.8	89.6
			28/0000	34.4	86.7

NOAA

4. Why do you think Andrew veered sharply north and east from 27/0000 to 28/0000?

Probability of Landfall

Table 7.2 contains data on the probability of landfall for Miami and New Orleans days prior to the actual event. Using the hurricane tracking chart you made earlier, and the graphs provided in Appendix B, plot probability on the *y*-axis against distance on the *x*-axis. Distance should be measured from your chart for the date and times provided in the table.

Table 7.2 Probability of Landfall

Miami		New Orleans	
Date/Time	Probability (%)	Date/Time	Probability (%)
		22/1200	3
21/1200	7	23/0000	7
22/0000	8	23/1200	12
22/1200	14	24/0000	12
23/0000	23	24/1200	21
23/1200	40	25/0000	21
24/0000	71	25/1200	36
25/0500	99	26/0000	66
		26/1200 (28N, 91W, landfall)	99

NOAA

5. Using your graphs, at what distance was the probability 50% for Miami?

6. At what distance was the probability 50% for New Orleans?

7. Is the relationship between probability and distance linear, or nonlinear?

Climate Change

EXERCISE 8

Climate is the long-term state of the atmosphere. Climate data is usually calculated as monthly or annual averages in temperature and precipitation. A common way to express climate data is by what are called normals. These are based on thirty-year averages, be it for any given date, month, or year. They are used for ten years and then recalculated over the last thirty years as a moving average.

1. Minneapolis/St. Paul Climate

Much debate has surrounded climate change, particularly global warming, over the last several decades. Interestingly, the debate was over global cooling in the 1970s and early 1980s. Several government studies in the 1990s indicate that the lower 48 states actually cooled slightly during the last century (Pioneer Press). This hypothesis will be tested in this exercise. Climate data for Minneapolis and St. Paul, Minnesota will be used from 1900 to 2000 (Appendix C).

Twentieth-Century Annual Average Temperature Trend

1. Using the graph paper provided in Appendix B, plot decade averages for the period starting with 1900–1909. Do this for each decade.

2. Which decade was the warmest?

3. Which decade was the coolest?

4. What is the range in temperature between the warmest and coolest decade?

Looking at the graph, it is difficult to see a temperature trend, either cooler or warmer, over the 100-year period. However, it is easy to plot a trend line on the graph. The simplest method is by the use of semi-averages. Calculate the average value for 1900–1949 decades. Plot that value on the graph for the 1920s, the middle decade for the first half of the series. Do the same for the 1950–1999 period using 1970s decade. Drawing a line through these two points should give you the trend.

5. Does your trend analysis confirm or refute the hypothesis of slightly cooler temperatures for the Twin Cities data during the twentieth century?

EXERCISE 9

Acid Precipitation Across the Nation

Acid precipitation has been recognized as a contributor to the damage of natural ecosystems for decades. Many lakes and forests have been irreparably damaged in Europe and the United States by highly acidic precipitation. The acidification of lakes, streams, and forests continue today. Figure 9.1 shows the most recent map of precipitation acidity in the United States.

The pH scale is a measure of hydrogen concentration in moles per liter. It is a negative log function. The higher the hydrogen concentration, the greater the acidity, and the lower the pH value. Neutral on the scale is 7.0. Values above 7.0 are considered basic and below 7.0 acidic.

Precipitation becomes acidified by the reaction of water or snow with nitrogen and sulfur oxides. Nitric, sulfurous, and sulfuric acids are rapidly formed in the atmosphere and fall as acid precipitation. Industry, power generation, and automobiles produce these oxides.

1. Which US regions have the lowest pH on the map in Figure 9.1?

2. What might account for the lower pH in those regions?

Rainwater has a natural pH of 5.6 because of the reaction of atmospheric moisture and carbon dioxide forming carbonic acid. A pH below 5.6 would be the result of other oxide reactions. Acid precipitation can be neutralized, or raised in pH, by certain compounds such as calcium

42 EXERCISE 9: ACID PRECIPITATION ACROSS THE NATION

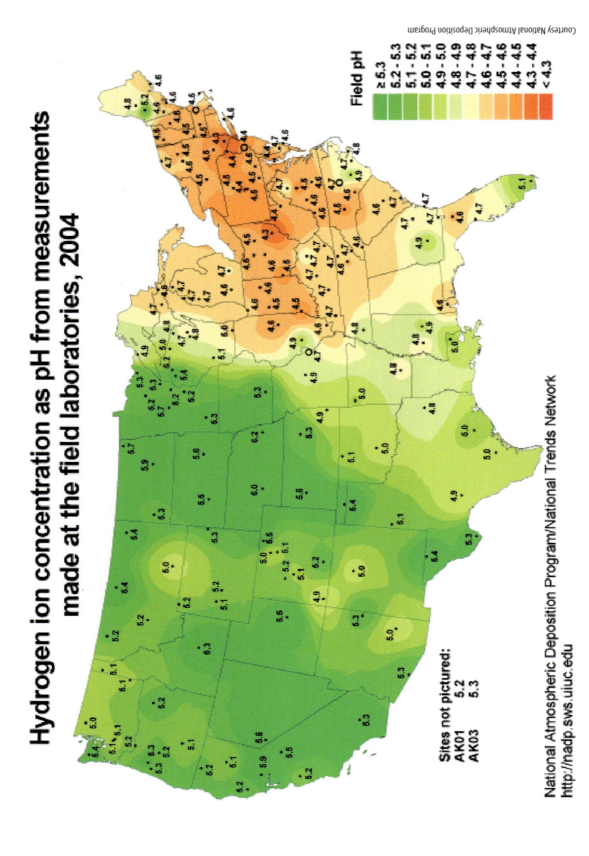

Figure 9.1 US Acid precipitation distribution represented as pH.

carbonate. This process is called buffering. Limestone, a common sedimentary rock, is made of calcium carbonate. Regions with limestone bedrock, or limey soils, are to a degree naturally protected.

3. Since monitoring began in the 1980s, the region with the highest pH has been the northern Great Plains, particularly the Dakotas. Average annual pH has often been higher than 5.6. What might account for this?

4. The Rocky Mountains evoke an image of fresh air and pristine waters. Notice however, from southern Montana to Arizona and New Mexico, precipitation pH is relatively low. Speculate on why this may be so.

EXERCISE 10

Topographic Map Interpretation

A topographic map is a general reference map created by the United States Geological Survey that incorporates the element of elevation throughout the map. It is extremely useful to many different professionals and policy makers. Several topographic maps will be used in the following exercises. Therefore, some mapping basics are needed to be reviewed for the student.

1. Direction

The top of any map is north unless otherwise indicated. From any point on a map, a compass direction of north, east, south, and west can be expressed as degree values corresponding to a 360 degree circle measured clockwise. Therefore, west corresponds to 270 degrees. A degree direction is called an azimuth. For example, an azimuth of 45 degrees would correspond to northeast. Any degree direction between two points can be drawn on a map using a protractor.

2. Scale and Distance

A map is a reduced version of the real world. This reduction is called the scale of the map. For example, a scale of 1–1,000 (expressed as 1:1,000) means that one unit measured on the map is equal to 1,000 of the same units in reality, be they inches, feet, or centimeters. The larger the number after the colon, the smaller the scale of the map. A wall map of the world might have a scale of 1:25,000,000, where one inch would equal 25,000,000 inches in reality. Small-scale maps show greater area and distance but at reduced detail. Large-scale maps show good detail but over a smaller area.

    A common scale for a topographic map is 1:24,000 which would be considered a relatively large-scale map. At this scale, one inch on the map would equal 24,000 in reality, or 2,000 feet. Don't forget that any unit can be used. For example, 2 centimeters on the map would equal 48,000 centimeters in reality, or 480 meters. Find the topographic map of Grand Marais, MN in Appendix A. Note the map scale of 1:24,000 and the bar scales below it. However, this map has been reduced to a scale of 1:28,000. Also notice all of the reference information on the map represented by the variety of map symbols. A key of topographic map symbols can be found in Appendix B.

3. Elevation

The real utility of a topographic map is its ability to represent elevation across the map. This can be done several different ways. Elevations can be written directly on the map. These are called spot elevations and are often found at road intersections and hill tops. A special type of spot elevation is called a benchmark, and will have a BM near the elevation on the map. There is an actual benchmark plate to be found at that point in the real world. The difference between the highest and lowest point on the map is called the relief of the map.

The method to express elevation across the entire map is by using lines of constant elevation called contour lines. Elevation along a contour line is constant. Contour lines cannot touch or cross because of this. Every fourth or fifth line is labeled. This is called an index contour for easier elevation orientation. The change in elevation from one line to another is called the contour interval, found at the bottom of the map. It is 20 feet on the Grand Marais map, but will vary from map to map, depending on the topography.

4. Contour Line Patterns

The pattern contour lines make can tell the interpreter a great deal. Lines close together indicate a rapid change of elevation over a short horizontal distance, or a steep slope, such as found near the Devil Track River. Notice how the contour lines form a V when crossing the river. The point of the V always points in the upstream direction whenever crossing a stream or river. Lines far apart indicate gentle slopes. Concentric circles, or lines within lines, indicate a hill or depression. A depression will have special marks called hachures on the contour line.

1. What is the azimuth from the gravel pit east of Chippewa City, to the benchmark (BM) south of the TV tower near the top of the map?

2. What is the total relief on the Grand Marais map?

3. What is the distance from the high school to the benchmark east of Chippewa City in feet? In miles? In kilometers?

The slope, or grade on a topographic map from one point to another, can be expressed as feet per mile, as a degree angle, or as percent grade. The percent grade is often used on highway signs in mountain regions or down steep river valleys. Percent grade between two points can be expressed by the difference in elevation, divided by the horizontal distance between two points, multiplied by 100. This is sometimes called the rise, divided by the run, times 100. Notice how the Gunflint Trail runs as parallel as possible to the contour lines to decrease the grade as much as possible.

4. a) What is the distance from the creek that crosses the Gunflint Trail near the word Trail, to Lake Superior?

   b) What is the change in elevation?

c) What is the approximate grade?

5. What is the approximate elevation change from the cemetery north of Chippewa City to the TV tower from question 1?

Regional Floods

EXERCISE 11

Floods killed more people during the 1990s in the United States than any other natural disaster. Most of these casualties occurred from flash floods in relatively small areas and short time frames.

Regional flooding on the other hand occurs over large areas, and may last for weeks or months. Fatalities are usually minimal because of sufficient warning, but displacement and destruction is often devastating. The Mississippi flood in 1993 was the worst in United States history, costing billions of dollars and affecting much of the mid United States (PBS).

Locate the Stillwater, MN topographic map in Appendix A. The scale of this map is 1:28,000 and a contour interval of 10 feet. This map shows the St. Croix River which forms part of the border between Minnesota and Wisconsin, 10 miles east of St. Paul. The town of Bayport, MN is built on a sandbar that is relatively flat, low lying, and prone to flooding. The normal river elevation is 675 feet. The largest flood crest occurred in the spring of 1965 at an elevation of 694 feet.

1. St. Croix River, Minnesota/Wisconsin Border

Locate the sewage disposal facility to the west of the St. Croix River label on the map. Note the 700 feet contour line on the map. Carefully highlight this line to the south for a reference elevation. The contour interval, the change in elevation between lines, is 10 feet on this map. A key of topographic map symbols can be found in Appendix A.

1. Using the 700 feet contour and the 690 feet contour closer to the river, how many structures would be inundated by a 694 feet flood crest?

2. The large building in the middle of the map is a window manufacturing plant employing hundreds of people in the area. How high would sandbags need to be to secure the plant for the flood? How high for the power plant to the north?

3. There are two churches near each other in the center of Bayport. Are they safe from a flood this size? What about the school a block to the west?

4. There is a marina near the center bottom of the map labeled Crocus Park. The buildings clustered to the immediate north are condominiums. How high would the floodwaters be up the exterior walls without sand bagging?

5. Why would industry and utilities locate in a flood prone area?

EXERCISE 12

Mega Disasters

1. Tsunami

Tsunami are ocean waves generated by water displacement caused by tectonic plate motion at oceanic subduction zones, volcanic eruptions, land and submarine slides, or asteroid impact. By far, most are caused by plate motion, with accompanying earthquakes. In 2004, a tsunami propagated through the Indian Ocean with devastating effects in Sumatra, Thailand, and Sri Lanka, killing hundreds of thousands.

In this exercise, the Pacific Ocean Basin will be examined for tsunami impact on a portion of the California coast just south of Los Angeles. Three actual tsunami source areas will be studied. Two of the tsunami were generated by subduction zone earthquakes near Alaska and Japan, and a third from an earthquake induced submarine landslide off the coast of Hawaii. Although, the loss of life from any of the three does not compare to the Sumatra disaster, deaths did occur, and waves 21 feet high did affect the California coast from the Alaskan event, and killed 12 people (Abbott).

1. Using the travel time maps for the 1964, 1968, and 1975 tsunami in Appendix A, determine the approximate azimuth from the initiation areas marked by the star, to Southern California.

2. Draw the azimuths calculated in question 1 on the topographic map of Redondo Beach in Appendix A from Malaga Cove. The scale of this map is 1:28,000 with a contour interval of 20 feet. Which tsunami wave front would have the most direct impact on Malaga Cove?

Tsunami is the Japanese word for harbor wave. The shape of the affected coastline and the slope of the seabed both have an influence on the resulting impact of a tsunami. Whereas a cove or a headland may shelter and protect an area on the coast from full impact, other areas may face the brunt of a direct approach of a tsunami. In addition, a shallow seabed over a long distance, approaching a shore, leads to greater wave amplification and height (NOVA).

3. Highlight the 80 feet contor line on the topographic map, as best you can, starting from the top of the map to the Malaga Cove area. Assume that a 60 feet tsunami affects the coast from all three azimuths. Would the Malaga Cove School survive any of the tsunami?

4. Most of the coast south of Malaga Cove should be protected by the headland cliffs. However, all three tsunami directions would affect much of the coast north of Malaga Cove. Note the many benchmarks along the shore. The scale of this map is 1:28,000, meaning that one unit on the map is equal to 28,000 units in reality. For example, one inch would equal 28,000 inches or 2330 feet. How long of a stretch would be affected north of Malaga cove?

5. How far inland would be the maximum distance of the tsunami?

6. Besides the relatively shallow gradient of the seabed, what other mitigating factors might contribute to loss of life?

2. Impact

In January 2002, an asteroid passed by Earth within 520,000 miles. It was estimated to be 1000 feet across and traveling about 68,000 mph. One scientist was quoted as saying if it had hit Earth it would take out a medium size country like France or something like Texas (Pioneer Press).

Objects from space have impacted the earth throughout history. One of the best examples is Meteor Crater in Arizona, formed 50,000 years ago. This meteor is estimated to have been 150 feet across, weighing 300,000 tons, and travelling at 28,600 mph. It struck the Earth with the force of 2.5 megatons of TNT (Barringer Crater).

This depression is classified as a simple crater, one that is circular and less than 3 miles in diameter (USGS). Nevertheless, debris was scattered over a 100 mile diameter area and would have extinguished all life within 2–3 miles. The resulting fireball would have caused flash burns up to 7 miles. The shockwave produced would have moved at 1200 mph and leveled everything within an 8.5–13.5 mile radius. Dissipating hurricane force winds would have continued up to 25 miles away (Trip Atlas).

Using the topographic map of Meteor Crater in Appendix A, answer the following questions. The scale of this map is 1:28,000 with a contour interval of 20 feet.

1. What is the approximate elevation of the craters rim? Approximately, how deep is the crater?

2. What is the approximate distance across the rim in meters? In feet?

3. Determine the average elevation in the vicinity of crater. How deep was the excavation disregarding overturned and uplifted bedrock, and debris that built the taller crater rim?

4. Examine the contours around the crater. Is there any pattern to be seen for ejected material from the impact? If so, can a direction of travel be determined for the meteor?

5. If people were living in the nearest town, Winslow, Arizona, would they have survived?

APPENDIX A
Maps and Charts

World map

Volcanic hazards map

Upper Midwest map

Atlantic hurricane chart

Grand Marais, MN topographic map

Stillwater, MN topographic map

Redondo Beach topographic map

Tsunami time travel maps

Meteor Crater, Arizona topographic map

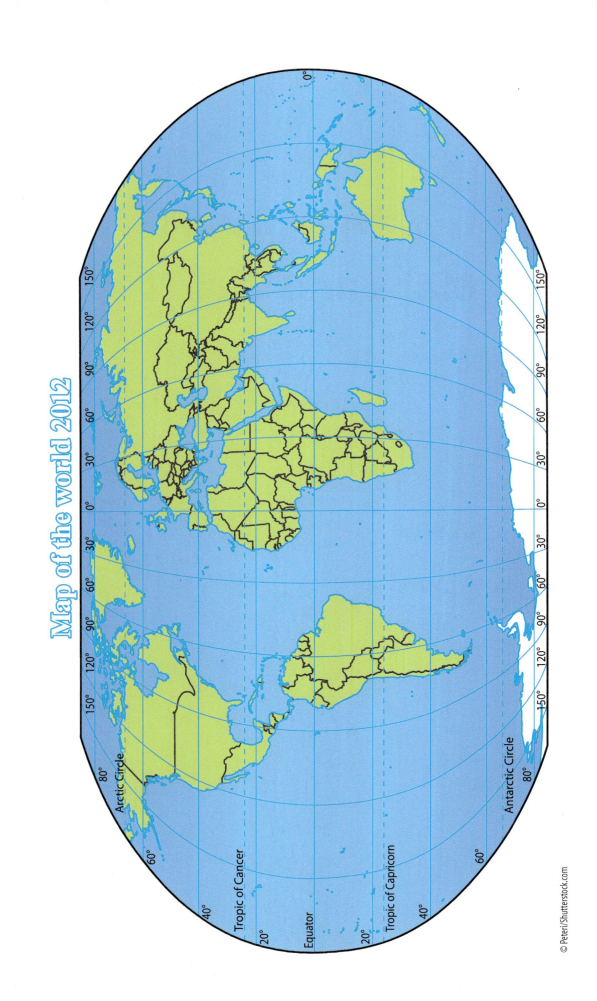

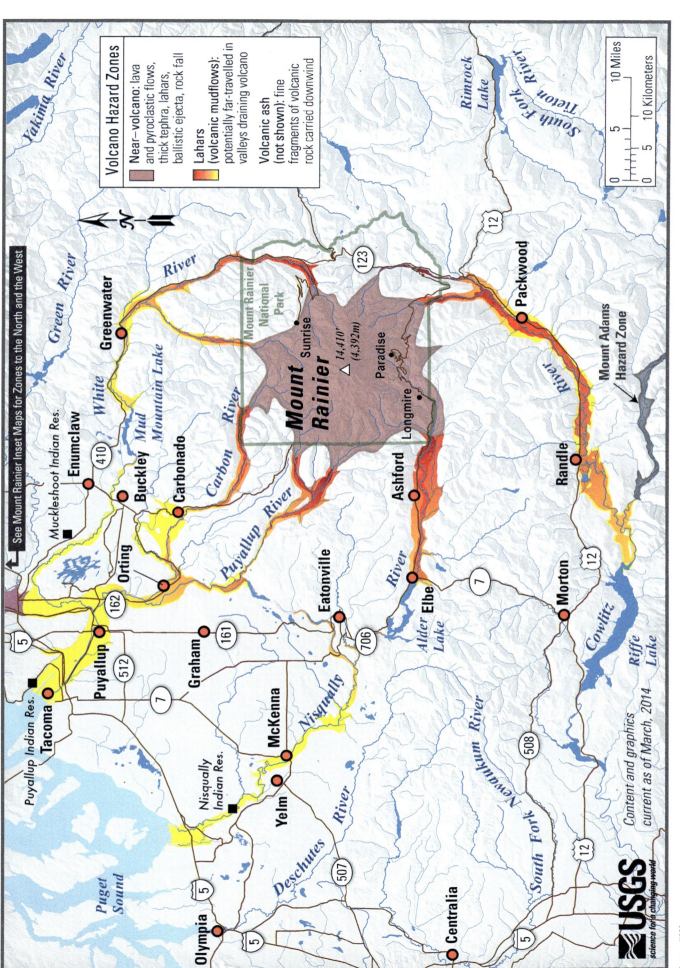

Courtesy USGS

Atlantic Basin Hurricane Tracking Chart
National Hurricane Center, Miami, Florida

Courtesy NOAA

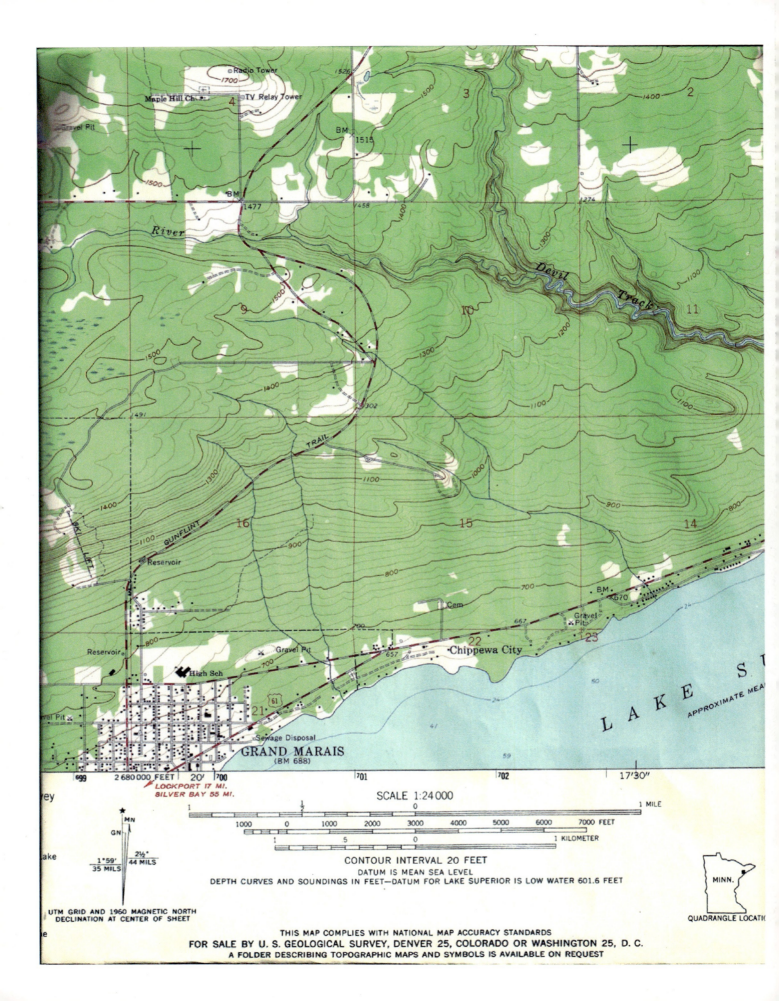

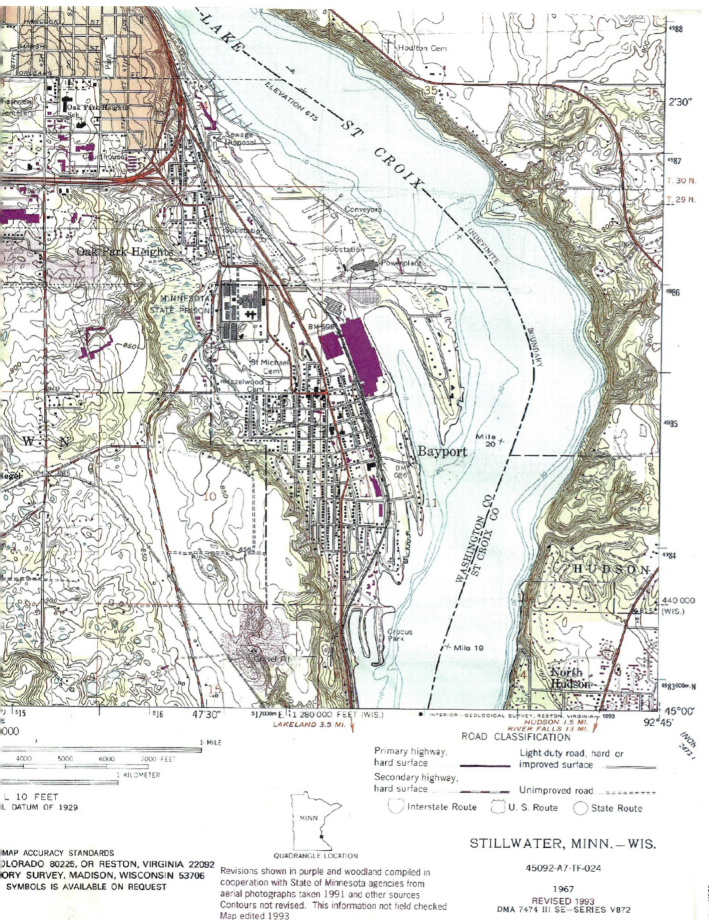

62 APPENDIX A: MAPS AND CHARTS

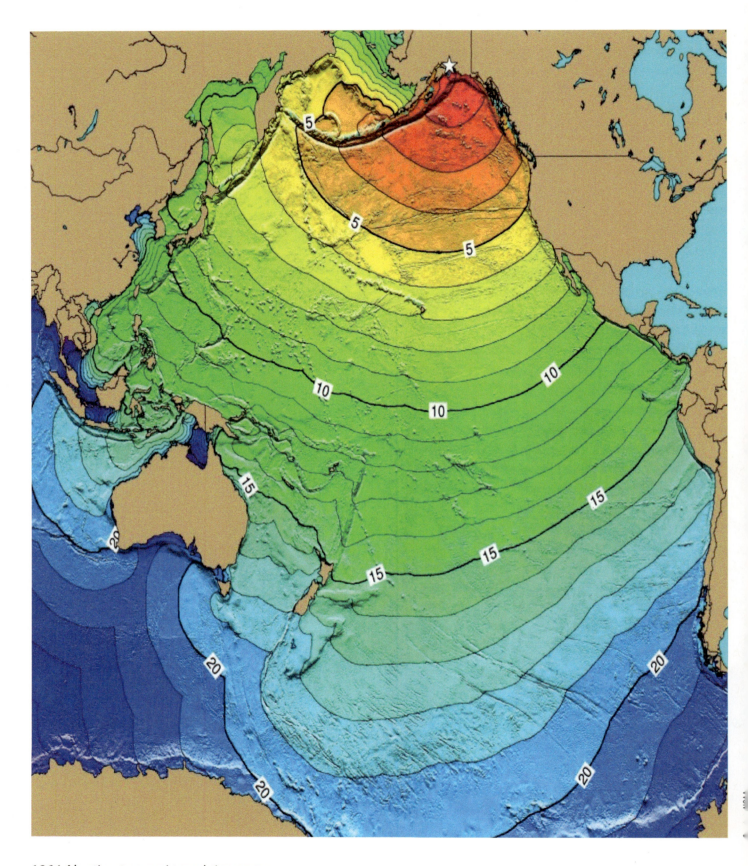

1964 Aleutian tsunami travel time map

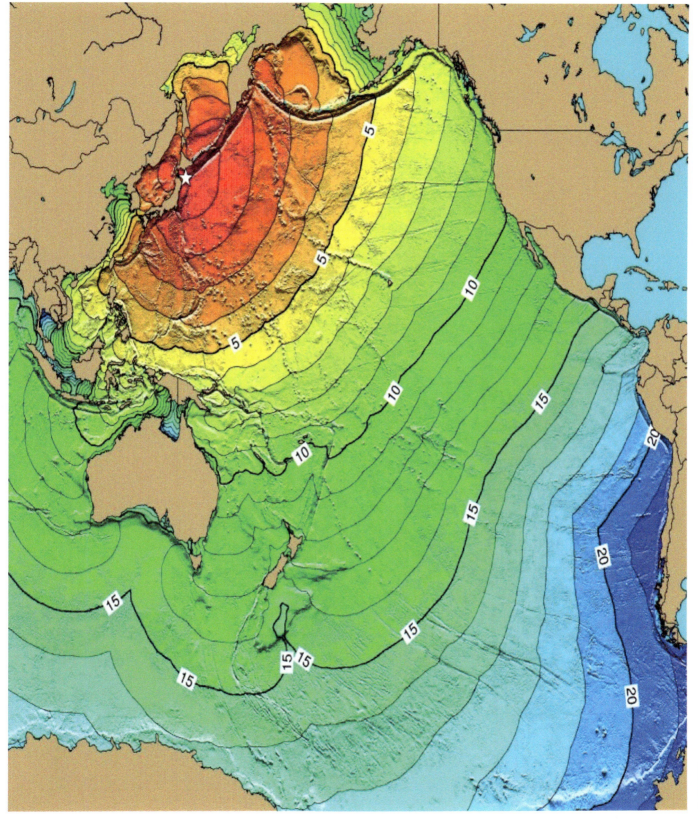

1968 Japan tsunami travel time tap

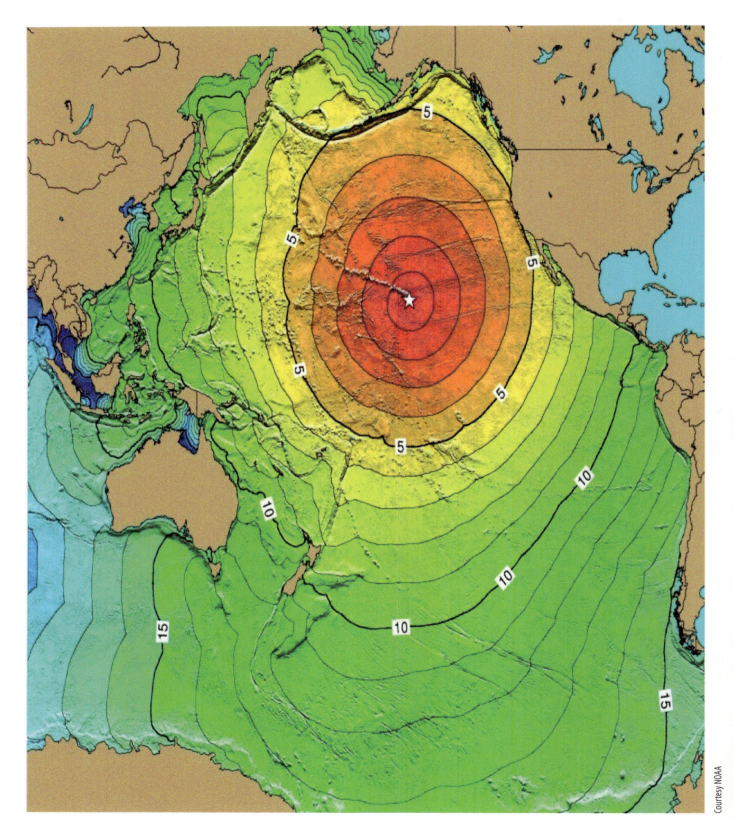

1975 Hawaiian tsunami travel time map

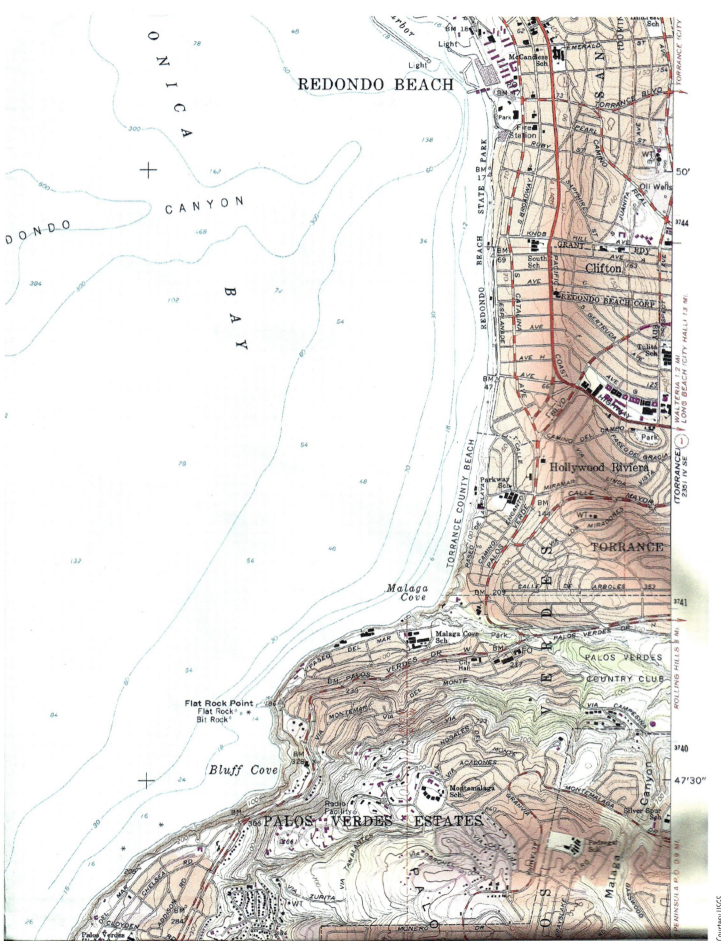

METEOR CRATER, ARIZ.
N3500—W11100/7.5

1968
PHOTOREVISED 1983

Revisions shown in purple compiled from aerial photographs taken 1980 and other source data.

ROAD CLASSIFICATION

| Heavy-duty | Light-duty |
| Medium-duty | Unimproved dirt ========= |

U.S. Route
Interstate Route

QUADRANGLE LOCATION: ARIZONA

Courtesy USGS

APPENDIX B
Analysis Tools

Graph paper

Conversion tables

Topographic map symbols

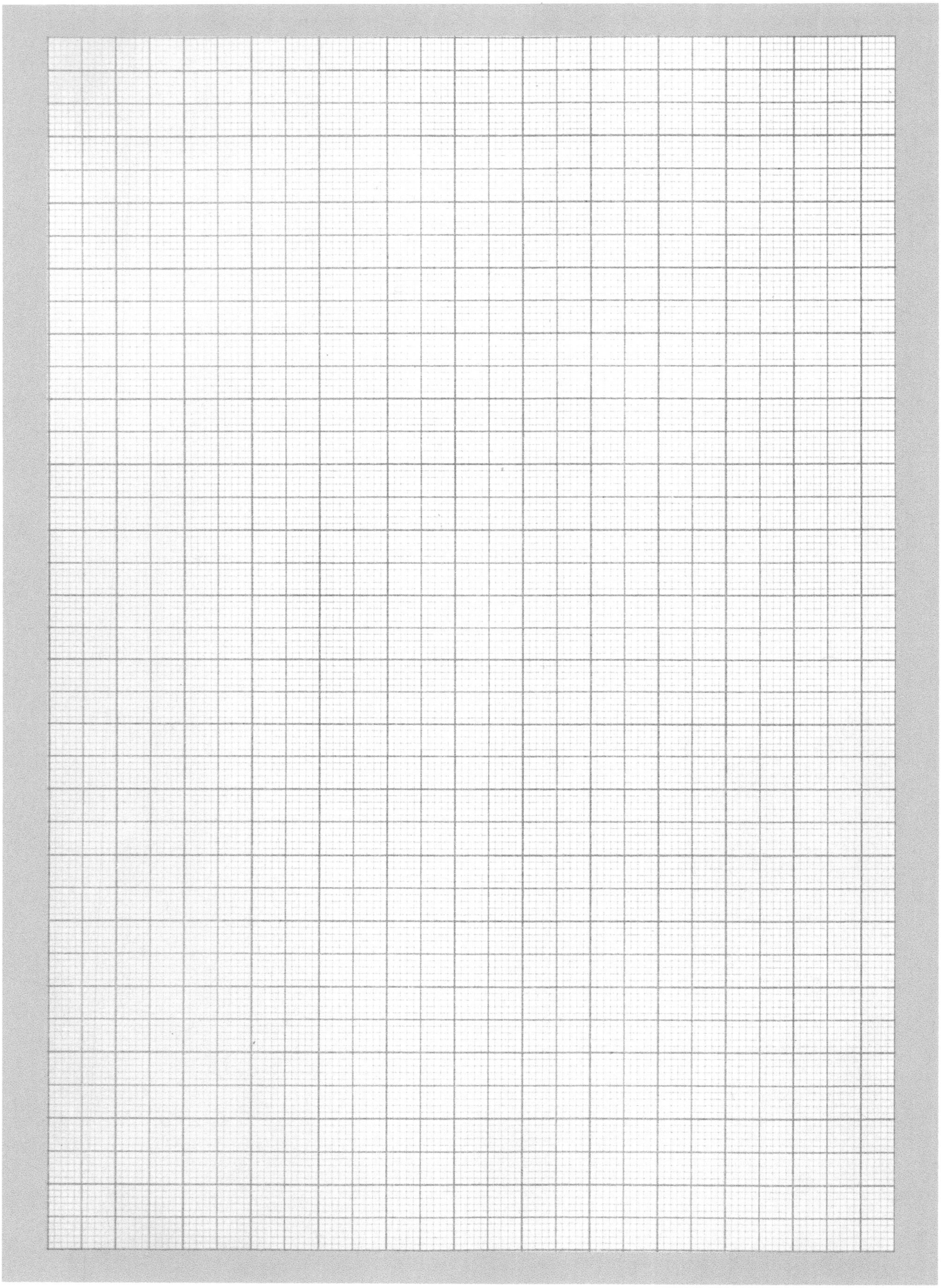

1. Conversion Formulas

Begin with	Formula for Converting (Multiply Number of Units by Conversion Number to Obtain New Number of Units)
millimeters (mm)	millimeters x 0.03937 = inches
centimeters (cm)	centimeters x 10 = millimeters centimeters x 0.3937 = inches
meters (m)	meters x 1000 = millimeters meters x 100 = centimeters meters x 3.281 = feet meters per second x 3.281 = feet per second square meters x 10.76 = square feet square meters x 1.196 = square yards square meters x 0.0002471 = square acres cubic meters x 35.31 = cubic feet cubic meters x 1.308 = cubic yards cubic meters x 0.0008107 = acre-feet cubic meters per second x 35.31 = cubic feet per second cubic meters per second x 15,850.00 = gallons per minute
kilometers (km)	kilometers x 1000 = meters kilometers x 0.6214 = miles square kilometers x 0.3861 = square miles cubic kilometers x 0.2399 = cubic miles
inches (in)	inches x 25.4 = millimeters inches x 2.54 = centimeters square inch x 6.4516 = square centimeters
feet (ft)	feet x 12 = inches feet x 0.3048 = meters square feet x 0.09294 = square meters cubic feet x 0.02832 = cubic meters acre-foot x 1233 = cubic meters
yard (yd)	yard x 3 = feet yard x 0.9144 = meters
miles (mi)	miles x 5280 = feet miles x 1609.3 = meters miles x 1.609 = kilometers square miles x 2.590 = square kilometers cubic miles x 4.168 = cubic kilometers

CONVERSION FORMULAS 71

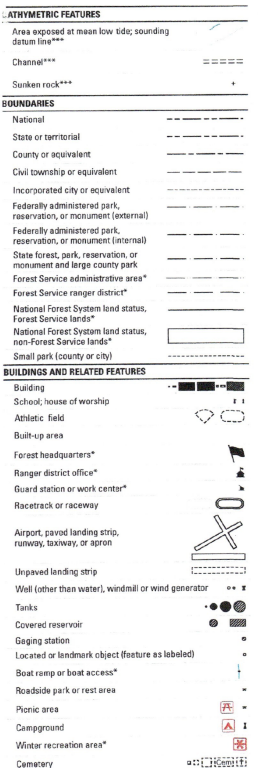

BATHYMETRIC FEATURES
- Area exposed at mean low tide; sounding datum line***
- Channel***
- Sunken rock***

BOUNDARIES
- National
- State or territorial
- County or equivalent
- Civil township or equivalent
- Incorporated city or equivalent
- Federally administered park, reservation, or monument (external)
- Federally administered park, reservation, or monument (internal)
- State forest, park, reservation, or monument and large county park
- Forest Service administrative area*
- Forest Service ranger district*
- National Forest System land status, Forest Service lands*
- National Forest System land status, non-Forest Service lands*
- Small park (county or city)

BUILDINGS AND RELATED FEATURES
- Building
- School; house of worship
- Athletic field
- Built-up area
- Forest headquarters*
- Ranger district office*
- Guard station or work center*
- Racetrack or raceway
- Airport, paved landing strip, runway, taxiway, or apron
- Unpaved landing strip
- Well (other than water), windmill or wind generator
- Tanks
- Covered reservoir
- Gaging station
- Located or landmark object (feature as labeled)
- Boat ramp or boat access*
- Roadside park or rest area
- Picnic area
- Campground
- Winter recreation area*
- Cemetery

COASTAL FEATURES
- Foreshore flat
- Coral or rock reef
- Rock, bare or awash; dangerous to navigation
- Group of rocks, bare or awash
- Exposed wreck
- Depth curve; sounding
- Breakwater, pier, jetty, or wharf
- Seawall
- Oil or gas well; platform

CONTOURS
Topographic
- Index
- Approximate or indefinite
- Intermediate
- Approximate or indefinite
- Supplementary
- Depression
- Cut
- Fill
- Continental divide

Bathymetric
- Index***
- Intermediate***
- Index primary***
- Primary***
- Supplementary***

CONTROL DATA AND MONUMENTS
- Principal point**
- U.S. mineral or location monument
- River mileage marker

Boundary monument
- Third-order or better elevation, with tablet
- Third-order or better elevation, recoverable mark, no tablet
- With number and elevation

Horizontal control
- Third-order or better, permanent mark
- With third-order or better elevation
- With checked spot elevation
- Coincident with found section corner
- Unmonumented**

Courtesy USGS

APPENDIX B: ANALYSIS TOOLS

CONTROL DATA AND MONUMENTS – continued

Vertical control

Third-order or better elevation, with tablet	BM ✕ 5280
Third-order or better elevation, recoverable mark, no tablet	✕ 528
Bench mark coincident with found section corner	BM + 5280
Spot elevation	✕ 7523

GLACIERS AND PERMANENT SNOWFIELDS

- Contours and limits
- Formlines
- Glacial advance
- Glacial retreat

LAND SURVEYS

Public land survey system

- Range or Township line
  - Location approximate
  - Location doubtful
  - Protracted
  - Protracted (AK 1:63,360-scale)
- Range or Township labels — R1E T2N
- Section line
  - Location approximate
  - Location doubtful
  - Protracted
  - Protracted (AK 1:63,360-scale)
- Section numbers — 1 - 36
- Found section corner
- Found closing corner
- Witness corner — WC
- Meander corner — MC
- Weak corner*

Other land surveys

- Range or Township line
- Section line
- Land grant, mining claim, donation land claim, or tract
- Land grant, homestead, mineral, or other special survey monument
- Fence or field lines

MARINE SHORELINES

- Shoreline
- Apparent (edge of vegetation)***
- Indefinite or unsurveyed

MINES AND CAVES

Quarry or open pit mine	✕
Gravel, sand, clay, or borrow pit	✕
Mine tunnel or cave entrance	⊸
Mine shaft	▪
Prospect	x
Tailings	Tailings
Mine dump	
Former disposal site or mine	

PROJECTION AND GRIDS

Neatline	39°15′ 90°37′30″
Graticule tick	55′
Graticule intersection	+
Datum shift tick	

State plane coordinate systems

Primary zone tick	640 000 FEET
Secondary zone tick	247 500 METERS
Tertiary zone tick	260 000 FEET
Quaternary zone tick	98 500 METERS
Quintary zone tick	320 000 FEET

Universal transverse mercator grid

UTM grid (full grid)	273
UTM grid ticks*	269

RAILROADS AND RELATED FEATURES

- Standard guage railroad, single track
- Standard guage railroad, multiple track
- Narrow guage railroad, single track
- Narrow guage railroad, multiple track
- Railroad siding
- Railroad in highway
- Railroad in road
- Railroad in light duty road*
- Railroad underpass; overpass
- Railroad bridge; drawbridge
- Railroad tunnel
- Railroad yard
- Railroad turntable; roundhouse

RIVERS, LAKES, AND CANALS

- Perennial stream
- Perennial river
- Intermittent stream
- Intermittent river
- Disappearing stream
- Falls, small
- Falls, large
- Rapids, small
- Rapids, large
- Masonry dam
- Dam with lock
- Dam carrying road

Courtesy USGS

CONVERSION FORMULAS 73

RIVERS, LAKES, AND CANALS – *continued*

- Perennial lake/pond
- Intermittent lake/pond
- Dry lake/pond
- Narrow wash
- Wide wash
- Canal, flume, or aqueduct with lock
- Elevated aqueduct, flume, or conduit
- Aqueduct tunnel
- Water well, geyser, fumarole, or mud pot
- Spring or seep

ROADS AND RELATED FEATURES

Please note: Roads on Provisional-edition maps are not classified as primary, secondary, or light duty. These roads are all classified as improved roads and are symbolized the same as light duty roads.

- Primary highway
- Secondary highway
- Light duty road
- Light duty road, paved*
- Light duty road, gravel*
- Light duty road, dirt*
- Light duty road, unspecified*
- Unimproved road
- Unimproved road*
- 4WD road
- 4WD road*
- Trail
- Highway or road with median strip
- Highway or road under construction
- Highway or road underpass; overpass
- Highway or road bridge; drawbridge
- Highway or road tunnel
- Road block, berm, or barrier*
- Gate on road*
- Trailhead*

* USGS-USDA Forest Service Single-Edition Quadrangle maps only.
In August 1993, the U.S. Geological Survey and the U.S. Department of Agriculture's Forest Service signed an Interagency Agreement to begin a single-edition joint mapping program. This agreement established the coordination for producing and maintaining single-edition primary series topographic maps for quadrangles containing National Forest System lands. The joint mapping program eliminates duplication of effort by the agencies and results in a more frequent revision cycle for quadrangles containing National Forests. Maps are revised on the basis of jointly developed standards and contain normal features mapped by the USGS, as well as additional features required for efficient management of National Forest System lands. Single-edition maps look slightly different but meet the content, accuracy, and quality criteria of other USGS products.

Printed on recycled paper

SUBMERGED AREAS AND BOGS

- Marsh or swamp
- Submerged marsh or swamp
- Wooded marsh or swamp
- Submerged wooded marsh or swamp
- Land subject to inundation

SURFACE FEATURES

- Levee
- Sand or mud
- Disturbed surface
- Gravel beach or glacial moraine
- Tailings pond

TRANSMISSION LINES AND PIPELINES

- Power transmission line; pole; tower
- Telephone line
- Aboveground pipeline
- Underground pipeline

VEGETATION

- Woodland
- Shrubland
- Orchard
- Vineyard
- Mangrove

** Provisional-Edition maps only.
Provisional-edition maps were established to expedite completion of the remaining large-scale topographic quadrangles of the conterminous United States. They contain essentially the same level of information as the standard series maps. This series can be easily recognized by the title "Provisional Edition" in the lower right-hand corner.

*** Topographic Bathymetric maps only.

Topographic Map Information
For more information about topographic maps produced by the USGS, please call:
1-888-ASK-USGS or visit us at http://ask.usgs.gov/

ISBN 0-607-96942-3

Courtesy USGS

APPENDIX C
Weather Information

Surface air pressure tutorial

Weather front symbols

Annual temperature data

APPENDIX C: WEATHER INFORMATION

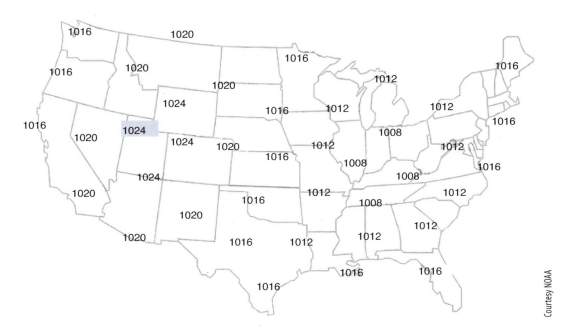

This map shows the sea level pressures for various locations over the contiguous United States. The values are in whole millibars.

1. Objective

Using a black colored pencil, lightly draw lines connecting identical values of sea level pressure. Remember, these lines, called isobars, do not cross each other. Isobars are usually drawn for every 4 millibars, using 1,000 millibars as the starting point. Therefore, these lines will have values of 1,000, 1,004, 1,008, 1,012, 1,016, 1,020, 1,024, etc., or 996, 992, 988, 984, 980, etc.

2. Procedure

Begin drawing from the 1,024 millibars station pressure over Salt Lake City, Utah (highlighted in blue). Draw a line to the next 1,024 value located to the northeast (upper right). Without lifting your pencil draw, a line to the next 1,024 value located to the south and then to the one located southwest, finally returning to the Salt Lake City value. Remember, isobars are smooth lines with few, if any, kinks.

The result is an elongated circle, centered approximately over Eastern Utah. The line that was drawn represents the 1,024 millibars line and you can expect the pressure to be 1,024 millibars everywhere along that line. Repeat the procedure with the next isobar value. Remember, the value between isobars is 4 millibars. Since there are no 1,028 millibars values on the map, then your next line will follow the 1,020 millibars reports. Then continue with the remaining values until you have all the reports connected with an isobar.

Label each isobar with the appropriate value. Traditionally, only the last two digits are used for labels. For example, the label on the 1,024 millibars isobar would be 24. A 1,008 millibars isobar would be labeled 08. A 992 millibars isobar will be labeled 92. These labels can be placed anywhere along the isobar but are typically placed around edges of the map at the end of each line. For closed isobars (lines that connect) a gap is placed in the isobar with the value inserted in the gap.

3. Analysis

Isobars can be used to identify "Highs" and "Lows". The pressure in a high is *greater* than the surrounding air. The pressure in a low is *lower* than the surrounding air.

- Label the center of the high pressure area with a large blue "**H**".
- Label the center of the high pressure area with a large red "**L**".

In addition to High and Low centers, you may see one or more of the following eight features on a surface analysis or forecast. The definitions provided below are derived from the National Weather Service Glossary.

	Cold Front—a zone separating two air masses, of which the cooler, denser mass is advancing and replacing the warmer.
	Warm Front—a transition zone between a mass of warm air and the cold air it is replacing.
	Stationary Front—a front between warm and cold air masses that is moving very slowly or not at all.
	Occluded Front—a composite of two fronts, formed as a cold front overtakes a warm or quasi-stationary front. Two types of occlusions can form depending on the relative coldness of the air behind the cold front to the air ahead of the warm or stationary front. A cold occlusion results when the coldest air is behind the cold front and a warm occlusion results when the coldest air is ahead of the warm front.
	Trough—an elongated area of relatively low atmospheric pressure; the opposite of a ridge. In the HPC's surface analyses, this feature is also used to depict outflow boundaries.
	Squall Line—a line of active thunderstorms, either continuous or with breaks, including contiguous precipitation areas resulting from the existence of the thunderstorms.

(dry line symbol)	**Dry Line**—a boundary separating moist and dry air masses. It typically lies north-south across the central and southern high Plains states during the spring and early summer, where it separates moist air from the Gulf of Mexico (to the east) and dry desert air from the southwestern states (to the west).
(curved line symbol)	**Tropical Wave**—a trough or cyclonic curvature maximum in the trade wind easterlies.

NOAA Satellite and Information Service
National Environmental Satellite, Data, and Information Service (NESDIS)

DOC > NOAA > NESDIS > NCDC   Search Field: Search NCDC

NCDC / Climate Monitoring / Climate At A Glance / Minneapolis-St.Paul / Search / Help

Minneapolis-St.Paul
Climate Summary
September 2005

The average temperature in September 2005 was 63.6 F. This was 4.5 F warmer than the 1895-2005 average, the 11th warmest September on record (1895-2005). The <u>non-adjusted</u> temperature in September 2005 was 66.3 F.

4.45 inches of precipitation fell in September. This was 1.54 inches more than the 1895-2005 average, the 19th wettest such month on record (1895-2005).

Select from the options below to view graphs and tables of monthly temperature and precipitation data for Minneapolis-St.Paul , then click "submit". (Please wait 20-30 seconds)

Data Type : Mean Temperature	First Year to Display : 1895
Period : Annual	Last Year to Display : 2005
Location: Minneapolis-St.Paul	⦿ Line Chart* ○ Bar Chart* ⦿ Table

Information for AOL users and others <u>experiencing problems</u> receiving requested output.

*A minimum of 8 years is required.

*

NCDC / Climate Monitoring / Climate At A Glance / Minneapolis-St.Paul / Search / Help

Privacy Policy   HOW ARE WE DOING? A user survey      Disclaimer

http://www.ncdc.noaa.gov/oa/climate/research/cag3/y7.html
Downloaded Tuesday, 01-Nov-2005 13:41:22 EST
Last Updated Monday, 31-Oct-2005 08:11:11 EST by Karin.L.Gleason@noaa.gov
Please see the NCDC Contact Page if you have questions or comments.

Climate At A Glance

Annual Temperature
Minneapolis-St.Paul, MN

(sorted by year)

Missing 1 year of data

Year	Temperature	Rank Based on the Time Period Selected (1895-2005)*	Rank Based on the Period of Record (1895-2005)*
2005	----	----	----
2004	43.9 deg F	83	83
2003	43.9 deg F	83	83
2002	44.5 deg F	93	93
2001	45.0 deg F	102	102
2000	43.9 deg F	83	83
1999	45.4 deg F	105	105
1998	46.3 deg F	109	109
1997	42.0 deg F	24	24
1996	39.9 deg F	3	3
1995	43.0 deg F	49	49
1994	43.0 deg F	49	49
1993	41.1 deg F	11	11
1992	42.9 deg F	43	43
1991	43.4 deg F	63	63
1990	44.8 deg F	99	99
1989	41.2 deg F	15	15
1988	44.0 deg F	85	85
1987	47.2 deg F	110	110
1986	43.2 deg F	51	51
1985	41.1 deg F	11	11
1984	42.8 deg F	40	40
1983	43.3 deg F	55	55
1982	42.1 deg F	25	25

http://climvis.ncdc.noaa.gov/cgi-bin/cag3/hr-display3.pl   11/1/2005

1981	43.9 deg F	83	83
1980	42.9 deg F	43	43
1979	41.0 deg F	9	9
1978	41.8 deg F	20	20
1977	42.9 deg F	43	43
1976	43.6 deg F	71	71
1975	42.6 deg F	36	36
1974	42.1 deg F	25	25
1973	44.4 deg F	90	90
1972	39.0 deg F	2	2
1971	41.8 deg F	20	20
1970	42.1 deg F	25	25
1969	42.4 deg F	33	33
1968	42.9 deg F	43	43
1967	40.4 deg F	6	6
1966	41.2 deg F	15	15
1965	40.5 deg F	7	7
1964	43.5 deg F	68	68
1963	42.8 deg F	40	40
1962	42.4 deg F	33	33
1961	43.6 deg F	71	71
1960	42.7 deg F	38	38
1959	43.8 deg F	76	76
1958	43.7 deg F	74	74
1957	43.2 deg F	51	51
1956	42.9 deg F	43	43
1955	43.3 deg F	55	55
1954	44.1 deg F	87	87
1953	44.8 deg F	99	99
1952	43.7 deg F	74	74
1951	39.9 deg F	3	3
1950	39.9 deg F	3	3
1949	44.5 deg F	93	93
1948	43.9 deg F	83	83
1947	43.4 deg F	63	63
1946	44.4 deg F	90	90
1945	41.8 deg F	20	20
1944	45.0 deg F	102	102

http://climvis.ncdc.noaa.gov/cgi-bin/cag3/hr-display3.pl

APPENDIX C: WEATHER INFORMATION

1943	42.1 deg F	25	25
1942	44.3 deg F	88	88
1941	45.9 deg F	107	107
1940	42.8 deg F	40	40
1939	44.9 deg F	100	100
1938	45.1 deg F	104	104
1937	42.5 deg F	35	35
1936	41.9 deg F	23	23
1935	43.4 deg F	63	63
1934	45.5 deg F	106	106
1933	44.8 deg F	99	99
1932	43.2 deg F	51	51
1931	49.0 deg F	111	111
1930	45.1 deg F	104	104
1929	41.1 deg F	11	11
1928	43.9 deg F	83	83
1927	42.1 deg F	25	25
1926	42.1 deg F	25	25
1925	43.5 deg F	68	68
1924	41.0 deg F	9	9
1923	43.9 deg F	83	83
1922	44.6 deg F	95	95
1921	46.2 deg F	108	108
1920	43.3 deg F	55	55
1919	42.7 deg F	38	38
1918	43.8 deg F	76	76
1917	38.9 deg F	1	1
1916	41.2 deg F	15	15
1915	43.3 deg F	55	55
1914	43.5 deg F	68	68
1913	44.0 deg F	85	85
1912	41.7 deg F	19	19
1911	43.5 deg F	68	68
1910	44.5 deg F	93	93
1909	42.2 deg F	31	31
1908	44.6 deg F	95	95
1907	41.1 deg F	11	11
1906	43.5 deg F	68	68

http://climvis.ncdc.noaa.gov/cgi-bin/cag3/hr-display3.pl   11/1/2005

Year	Temp	Rank	Rank
1905	42.3 deg F	32	32
1904	40.6 deg F	8	8
1903	41.6 deg F	18	18
1902	43.4 deg F	63	63
1901	43.7 deg F	74	74
1900	44.8 deg F	99	99
1899	42.9 deg F	43	43
1898	44.1 deg F	87	87
1897	42.6 deg F	36	36
1896	43.6 deg F	71	71
1895	43.2 deg F	51	51

*Highest temperature rank denotes the hottest year for the period.
Lowest temperature rank denotes the coldest year for the period.

NCDC / Climate At A Glance / Climate Monitoring / Search / Help

This table was dynamically generated 11/01/2005 at 13:42:31
via http://www.ncdc.noaa.gov/oa/climate/research/cag3/cag3.html
Please send questions to Karin.L.Gleason@noaa.gov
Please see the NCDC Contact Page if you have questions or comments.

CREDITS

EXERCISE 1
Figure 1.1, http://nationalatlas.gov./articles/mapping/a_latlong.html

EXERCISE 2
Figure 2.1, http://science.nasa.gov/science-news/news-at-nasa/2000/ast06_1/
Figure 2.2, http://vulcan.wr.gov/Glossary/PlateTectonics/Maps/map juan de fuca ridge.html
Figure 2.3, http://hvo.wr.gov/volcanoes/

EXERCISE 3
Table 3.1, http://earthquake.usgs.gov/earthquakes/eqinthenews/2009/

EXERCISE 4
Figure 4.1, http://earthquake.usgs.gov/learn/glossary/?termID=149
Figures 4.2–4, http://www.ncedc.org/cgi-bin/make_seismogram.pl
Figure 4.5, http://earthquake.usgs.gov/learn/kids/eqscience.php
Figure 4.6, http://seismo.berkeley.edu/bdsn/bdsn_map.html

EXERCISE 5
Figure 5.1, http://scienceblogs.com/starts with a bang/2009/08/why_our_analemma_looks_like_a.php
Figure 5.2, http://www.srh.noaa.gov/abq/?n=clifeatures winter solstice

EXERCISE 6
Figure 6.1, http://www.hpc.ncep.noaa.gov/dailywxmap
Figure 6.2, http://www.hpc.ncep.noaa.gov/html/fntcodes_printer.html
Figure 6.3–6.4, http://www.hpc.ncep.noaa.gov/html/stationplot_printer.html
Table 6.1, http://www.hpc.ncep.noaa.gov/dailywxmap

EXERCISE 7
Table 7.1–7.2, html://wwwnhc.noaa.gov/1992andrew.html

EXERCISE 8
St. Paul Pioneer Press, May 24, 1990; April 20, 1999.

EXERCISE 9
Figure 9.1, http://nadp.sws.uiuc.edu

EXERCISE 12
Abbott, Natural Disasters 7th Ed. McGraw Hill.
Nova, The Day That Changed the World video.
St. Paul Pioneer Press, January 9, 2002.
http://arizona.usgs.gov/Outreach/…3CraterMorphology/cratermorph.pdf
www.barringercrater.com/science/main.htm
http://tripatlas.com/Barringer_Crater

APPENDIX A
World map, University of Minnesota Department of Geography.
Volcanic Hazards from Mt. Rainier, http://vulcan.wr.usgs.gov/Volcanoes/Rainier/Publications/FSO65-97_map.pdf
Upper Midwest map, source unknown.
Atlantic Basin Hurricane Tracking Chart, http://nhc.noaa.gov/tracking_charts.shtml
USGS Grand Marais, Minn. topographic quadrangle, 1960, photo revised 1986.
USGS Stillwater, Minn.-Wis. topographic quadrangle, 1967, revised 1993.
USGS Redondo Beach, CA. topographic quadrangle 1996.
Tsunami travel time maps, National Geophysical Data Center.
http://ngdc.noaa.gov/hazard/tsu_travel_time.shtml

APPENDIX B
Conversion formulas, http://vulcan.wr.usgs.gov/Miscellaneous/Conversion Tables/conversion_table.html
Topographic map symbols, http://egsc.usgs.gov/isb/pubs/booklets/symbols/Topomapsymbols.pdf

APPENDIX C
Surface air pressure tutorial, http://www.svh.noaa.gov/jetstream/synoptic/ll_analyze_slp.html
Weather front symbols, www.hpc.noaa.gov/html/fntcodes_printer.html
Annual temperature data, http://www.ncdc.noaa.gov/oa/climate/research/cag3/y7.html